MONKEYS
OF THE
AMAZON

First published in the United Kingdom in 2007 by:
Evans Mitchell Books
The Old Forge, Forge Mews
16 Church Street
Rickmansworth
Hertfordshire WD3 1DH
United Kingdom
www.embooks.co.uk

Jacket and Book Design by:
Roy Platten
Eclipse
Roy.eclipse@btopenworld.com

British Library Cataloguing in Publication Data.
A CIP record of this book is available
on request from the British Library.

ISBN: 978-1-901268-10-2

Pre Press: F E Burman, London, United Kingdom

Printed in Thailand

MONKEYS
OF THE
AMAZON

NICK GORDON

Evans Mitchell Books

Contents

Left: A close-up of a male spider
monkey's face – filled with
character and intelligence.

Introduction

I was hiding in a cramped, damp, dark film hide 35m (115ft) up in the roof of the forest and it was dawn. I had been there all night; my legs and back ached and I yawned. The creature resting in the fork of a branch 10m (33ft) in front of me was almost human-like. Its limbs, movements and mannerisms were all unsettlingly familiar to me; and then it yawned too.

It was its eyes that really held my attention. My face and camera were hidden by swathes of camouflage material, so surely it couldn't see me? Yet, those eyes were staring straight at me, and were filled with intelligence and suspicion.

In Amazonia, there are at least 118 species of primates known to science. Further, if you take into account the sub-species known or recognised at the present time, the total is a staggering 209. This is by far the highest primate diversity of any region in the world; yet even today, new species are still being discovered.

Right: The satere marmoset was only discovered a few years ago. It has been named after the Indians who live in the same area of forest.

Above: The latest species discovery, a dwarf marmoset, found in a tiny area of forest south of the Amazon River.

Being in the Amazon forest and knowing that species of the greatest diversity of primates in the world surrounds you is one thing; seeing them is quite another matter. They lead secret lives, hidden from human view for the most part, high up in the thick, tangled vegetation, 30m (over 100ft) or more above ground.

By 1994, I had lived in Amazonia for four years and had made three, one-hour television specials. Those films had focussed on the lives of giant otters in Guyana, huge tarantulas in Venezuela and the aquatic wildlife of an exceptional, remote area of northern Brazil. This forest traverses the political borders of eight countries. What had become clear to me from those early years was how the existence of almost every creature and plant that lived here depended on each other to survive. I wanted to make a film that would tell that story. However, at first, it seemed a hopeless and impossible task – that is, until I came to know the spider monkey.

I did not then realise that, over the next five years, I would come to know, understand and record on film for the first time, the secrets of many of these amazing Amazon monkeys.

Right: An adult female black spider monkey getting drunk on fermenting fruit, the long fingered hands are thumbless.

Distribution

The Amazon rainforest is home to more species of primates than anywhere else in the world. There are no apes (tail-less primates), just monkeys, marmosets and tamarins. The latter two types are seen as more primitive: they have claws on all digits except one and do not have prehensile tails. Otherwise, all Amazon monkeys have nails on all digits and can be broadly divided into those with or without prehensile tails.

One single factor affects the geographical distribution of Amazon monkeys more than any other – the rivers. Monkeys cannot swim, so these natural barriers have played a large part in forming the physical division and diversity between all of the primates here. This 'riverine barrier' hypothesis of distribution was first put forward by the great naturalist Alfred Wallace in 1852.

Of course, there are occasional circumstances where, due to some extraordinary event, a species has crossed over these borders. Perhaps a river that has dried up in one season gives those that visit the ground the rare opportunity to move to another area. It is also conceivable that when

Above: My first view of a wild satere marmoset – a female.

Left: High up in the canopy, a female woolly monkey with her three-month-old baby clinging on tightly.

Overleaf, right: The wonderfully expressive face of a male Satere marmoset.

Distribution maps – Monkeys of the Amazon

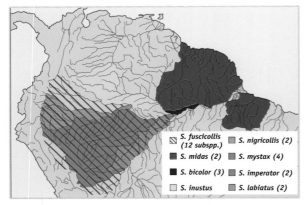

Legend:
- S. fuscicollis (12 subspp.)
- S. midas (2)
- S. bicolor (3)
- S. inustus
- S. nigricollis (2)
- S. mystax (4)
- S. imperator (2)
- S. labiatus (2)

Tamarins

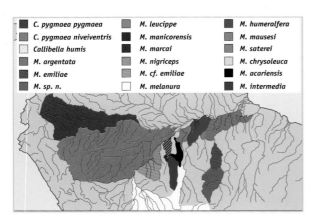

Legend:
- C. pygmaea pygmaea
- C. pygmaea niveiventris
- Callibella humis
- M. argentata
- M. emiliae
- M. sp. n.
- M. leucippe
- M. manicorensis
- M. marcai
- M. nigriceps
- M. cf. emiliae
- M. melanura
- M. humeralfera
- M. mausesi
- M. saterei
- M. chrysoleuca
- M. acariensis
- M. intermedia

Marmosets

Location of Amazonia

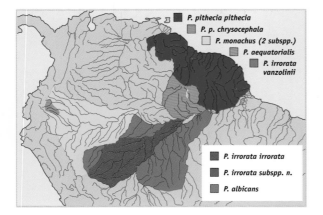

Legend:
- P. pithecia pithecia
- P. p. chrysocephala
- P. monachus (2 subspp.)
- P. aequatorialis
- P. irrorata vanzolinii
- P. irrorata irrorata
- P. irrorata subspp. n.
- P. albicans

Saki Monkeys

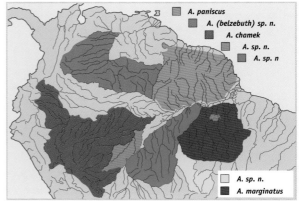

Legend:
- A. paniscus
- A. (belzebuth) sp. n.
- A. chamek
- A. sp. n.
- A. sp. n
- A. sp. n.
- A. marginatus

Spider Monkeys

Above: A female spider monkey with her year-old infant.

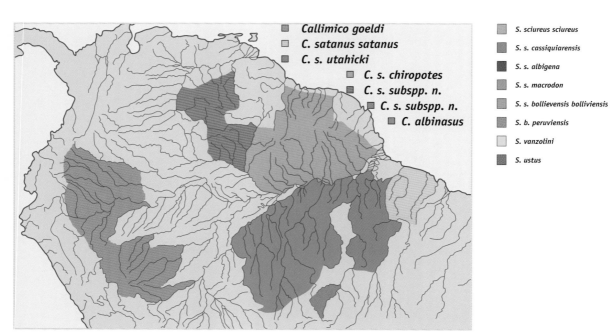

Goeldi's Monkey and Bearded Sakis

Legend:
- **Callimico goeldi**
- **C. satanus satanus**
- **C. s. utahicki**
- **C. s. chiropotes**
- **C. s. subspp. n.**
- **C. s. subspp. n.**
- **C. albinasus**

- *S. sciureus sciureus*
- *S. s. cassiquiarensis*
- *S. s. albigena*
- *S. s. macrodon*
- *S. s. bollievensis bolliviensis*
- *S. b. peruviensis*
- *S. vanzolini*
- *S. ustus*

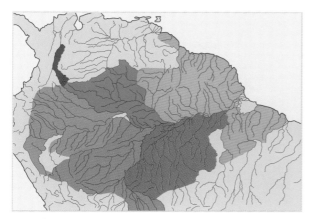

Goeldi's Monkey and Bearded Sakis

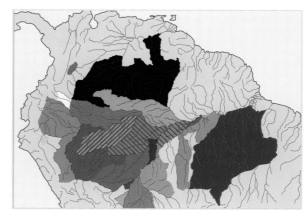

Titi Monkeys

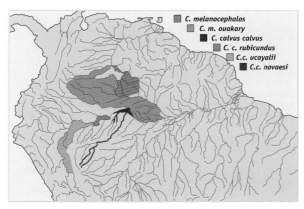

Uakari Monkeys

■	*C. melanocephalos*
■	*C. m. ouakary*
■	*C. calvus calvus*
■	*C. c. rubicundus*
■	*C.c. ucayalii*
■	*C.c. novaesi*

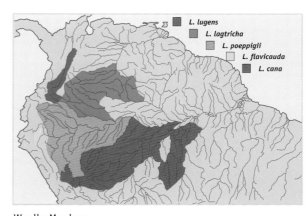

Woolly Monkeys

■	*L. lugens*
■	*L. lagtricha*
■	*L. poeppigii*
■	*L. flavicauda*
■	*L. cana*

a huge tree falls across a watercourse, or even floats down river, it just may have a monkey or two on board. Nevertheless, for the most part, the water presents an absolute barrier to primate movement.

This circumstance is never more apparent than within the small monkey species. The map of the area south of the central Amazon River shows clearly how the many small rivers separate the individual species. In these confines, over time, they evolve different physical characteristics. In some cases the areas in which just one type of monkey exists are truly tiny by Amazon standards.

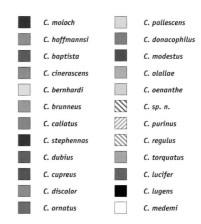

■	*C. moloch*	□	*C. pallescens*
■	*C. hoffmannsi*	▨	*C. donacophilus*
■	*C. baptista*	■	*C. modestus*
■	*C. cinerascens*	■	*C. olallae*
□	*C. bernhardi*	▨	*C. oenanthe*
■	*C. brunneus*	▨	*C. sp. n.*
■	*C. caliatus*	▨	*C. purinus*
■	*C. stephennas*	▨	*C. regulus*
■	*C. dubius*	■	*C. torquatus*
■	*C. cupreus*	■	*C. lucifer*
■	*C. discolor*	■	*C. lugens*
■	*C. ornatus*	□	*C. medemi*

Key applies to maps on pages 12, 14 and 15.

Habitat

From one's first steps into the Amazon rainforest, it seems that in terms of vegetation, chaos reigns. What is not immediately apparent is the way in which the forest is structured. But after a while, it becomes quite clear that, from the forest floor to the very tops of the trees, there are defined zones.

The lower storey, where huge trunks climb to 30m (98ft) or more, also contains smaller trees and saplings. These grow very slowly, waiting for one of the old giants to fall and expose a gap in the canopy, which each then tries to fill rapidly, before the competition gets there to monopolise the sunlight.

The middle layer is where the crowns of smaller trees fan out and many palm trees stand. Just above them, are the lower branches and boughs of the giant trees, some of which are festooned with epiphytes – plants that grow on other plants rather than in soil. These include species such as the bromeliads, which live on the nutrients they can snatch from their surroundings. Much of this middle zone interlinks and forms what are effectively highways and roads for the many primates that live here.

Above: Millions of square kilometres, covered with trees is perfect terrain for many monkeys.

Right:: An eagle's eye view of the great Amazon River. It is this river and its thousands of tributaries that have shaped the distribution of monkey species in this forest.

16

Left: An adult male woolly monkey patrols his territory high in the canopy.

Above: The giant crown of an emergent Parkia pendula tree.

Right: A single flowering tree. Its nearest relative is kilometres away and this is why some monkeys have to travel considerable distances to find food sources – this is biodiversity.

The five layers of the Tropical Forest

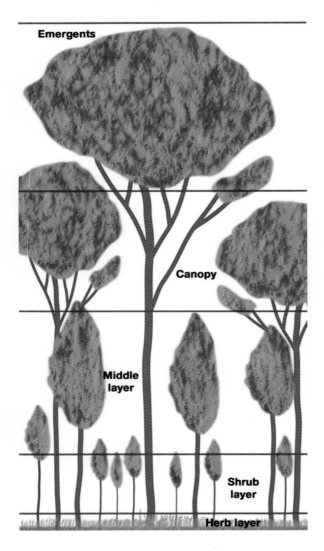

At the top are the **emergents**, those scattered trees that rise above the general canopy. Reaching 35m+ (120ft), the emergents are frequented by such as woolly monkeys.

Next comes the **canopy** layer that forms an almost continuous surface at around 25m (80ft) above the ground. This is home to most monkey families during the daytime when sunlight keeps the temperatures high. At night, however, heat loss by radiation is significant.

Below the canopy, reaching its lower boughs at around 15m (50ft), is the **middle** layer. From this level down, the ranges of temperatures enjoyed by the top two layers, which are in direct sunlight, are much smaller.

Saturated with moisture most of the time, the next level is that of woody and herbaceous **shrubs** that rise to 2–5m (6–15ft).

Finally, at ground level, the **herb** layer reaches up to a metre (2–3ft) in height.

For monkeys, the key to a harmonious existence is to maintain separation between the primate groups. This is achieved by claiming territory. Territories are defended and protected to varying degrees and all with one real aim – to protect the valuable and finite food source within these boundaries.

Trees and primates have evolved together over aeons, during which different monkeys have learned to exploit different foods in the various forest layers, thereby minimising contact and conflict with others of their kind.

Right: Clouds rising up from the rainforest.

Above left: A white uakari stares at the camera from the giant limb of a canopy tree.

Above right: A white-nosed bearded saki.

Left: Many plants and trees protect their precious fruits by providing homes for aggressive ants.

Right: Epiphytic plants scramble over tree trunks and branches providing refuge for many insect species, excellent hunting grounds for marmosets.

It is simply remarkable how these strategies play out. Large monkeys – such as the spider monkeys – are, in some cases, the exclusive feeders on a particular tree species. It is important to realise that the tree itself has also played an active role in this evolution. The plant's way of ensuring its continued survival is to produce abundant and viable seed; and it is vital that some creature is attracted to it, persuaded to remove the seed, take it away somewhere far enough from the parent tree, drop it and thereby allow it to produce the next generation. The fleshy sweet fruit, of course, is the payment to the monkey for this service.

Take strychnos as an example, this is a huge vine that is the source of the deadly poison strychnine. The Strychnos seed is laced with the poison and to chew the seed is certain death – I have seen the bodies of rodents like pacas lying on the forest floor among the fallen fruits. The spider monkey, however, does not masticate (chew) its food; instead, it swallows the fruits whole. The monkey's digestive system removes the nutritious flesh and the seed is eventually passed in their droppings – also useful as a free packet of fertiliser! This usually happens a considerable distance away from the original plant and thereby maintains the biodiversity of the area.

Right: Marmosets, such as this black and white tassel-ear, are perfectly adapted to life in the lower level of the forest and even visit the forest floor occasionally.

Primates have learned how to exploit the various food sources as they appear at different times of the year and in the different layers of the forest. The tiny marmosets and tamarins rarely visit the dizzy heights of the jungle canopy and, instead, have their territories in the lower forest layer.

Right: Marmosets, like this golden white tassel ear, use tree holes in the middle forest layer as sleeping dens at night.

Left and below: The small lightweight monkeys such as marmosets and tamarins can use the thinner branches and even foliage like palm fronds as walkways between trees.

Here, they take advantage of the fruits and insects and, for the marmosets in particular, there is another abundant food that others cannot generally get to – tree sap. Amazingly, these monkeys have evolved chisel-like lower teeth that allow them to gouge holes in the smaller trunks, where they feed on the tree resins that flow from the pits.

Inevitably, the rainforest is a complex and fragile ecosystem that is easily damaged. Left in a natural state, it flourishes, but remove just one element and the result can be catastrophic.

Right: Bromeliads, which include the pineapple, provide a source of water and food.

Below: Guarana seeds contain seven times more caffeine than coffee beans and are now an important commercial export from the forest.

Above left: Dugetia fruit grows directly off the side of the tree trunk.

Above right: Astracarium palm flowers attracting bees.

Left: Flowering passion fruit and euglossine bees.

In simple terms, chop the trees down and the monkeys' food has disappeared so they need to move to another area, if indeed there is one. Conversely, remove a certain species of monkey, by over hunting, say, and as sure as night follows day, the tree that depends upon it will also disappear. Yet the consequences are not immediately apparent, because a tree's life span is considerably longer than our own – and that is why most humans do not grasp the seriousness of the situation. In the long term, that particular tree species is condemned because there is no one to disperse its seeds.

Above: Fruit even grows in the lowest forest level, this grape-sized red fruit designed to attract creatures with colour vision – like the small monkeys.

Below: A seedling germinating on the surface of a tiny branch.

Right: An adult female golden white marmoset gouging a hole in a tree trunk to feed on resin.

Distinguishing Features

As we have seen, there are a great many species of New World primates. These are often small groups that have evolved apart from others due to the riverine barriers. Not all of these live in the Amazonian rainforest, although most do. As is immediately obvious, there are too many species to discuss and illustrate in detail in this book. I have tried to pick out some of those that I consider to be the more interesting.

The Amazon has an astonishing number of monkeys which fall into three main groups: large monkeys (pages 34–47), marmosets – including the world's smallest monkey, the pygmy marmoset – (pages 48–59) – and tamarins (pages 60–65).

Above: Their thick, dense, woolly fur, after which they were named, helps them keep warm during the cold Amazon nights.

Right: At dawn, a pair of spider monkeys hang from an upper canopy branch.

Monkeys

In spite of their numbers, monkeys are quite difficult to spot in the rainforest: in some areas they are common but their activities rarely bring them within sight or sound of humans. In areas where they have been hunted, they are extremely wary and disappear into the forest before you even realise that they are there.

Spider monkeys are more slender than their close relatives, the woolly monkeys. They have no thumbs, but make great use of their prehensile (grasping) tails when on the move in the trees.

Above: A female spider monkey still carries her year-old infant.

Right: A recently-discovered, new species of spider monkey.

Woolly monkeys also have prehensile tails that are used to wonderful effect in the high branches, where they spend most of their time. This tail has a bare patch at the tip of its underside that actually includes a thumbprint. Found throughout the Amazon, the woolly monkey gets its name – and this is obvious when you see it – from its thick, coarse fur.

Spiders and woollys can claim the 'swingers trophy', but the most vocal of the large monkeys is the aptly named howler monkey. Specialised vocal chords in the throat produce a howl that can be heard several kilometres away. A further difference is that howlers survive on a diet very different to that of their more acrobatic relations.

A specialised digestive system allows them to eat huge quantities of leaves which then ferment slowly while the howlers are at rest in the highest branches.

Above: An adult black spider monkey foraging for a last meal at dusk, the strongly prehensile tail is used expertly as a fifth limb.

Left: Adult female yawning at dawn; the canopy sunrise is reflected in her eyes.

Above: 'I was hidden in my film hide, yet her eyes seemed to be looking right into my lens...' Female adult in canopy.

Right: A juvenile alarm-calling after a harpy eagle flew by.

Opposite page: An adult male reaches out for succulent new growth, the large stomach aids the long digestion period needed for a leafy diet.

Overleaf, left: Alert to every sound, this female woolly monkey listens for the alarm calls of birds, which may warn of approaching predators.

Overleaf, right: A juvenile woolly monkey in the canopy.

The titis, sakis and uakaris are all found in the Amazon basin. The small titis show some similarities in social life to the tamarins and marmosets in that the male cares for the offspring. Active during the day, they eat mainly unripe fruits and insects high up in the forest canopy. Pairs mate for life and offspring stay until about three years old, The resulting family groups sleep in thickets of vines with their tails entwined in a charming fashion.

Above: The diminutive Squirrel monkey is seen frequently in large groups along river edges.

Right: The white uakari is known locally as the English monkey because of its sunburnt-looking face.

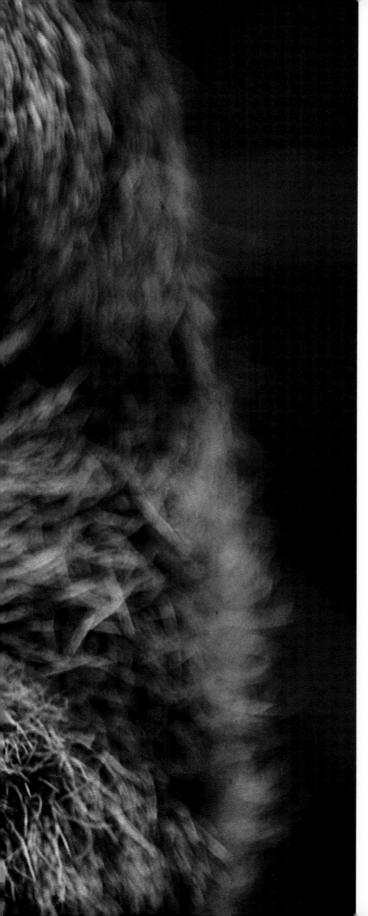

The short-tailed uakaris live in groups whose size ranges from five to 30 individuals. Their communication includes vocalisations and facial expressions, but their most obvious distinguishing feature is the difference between their shaggy coats and their bald heads and red faces. Uakaris bald with age much more completely than humans, and, as with humans, male uakaris generally lose more hair and become more bald than females.

The sakis mate for life and live in small family groups of parents and offspring. They are mainly fruit-eating although they will catch small mammals, including bats, and birds. They are almost as athletic as the capuchins and squirrel monkeys, but without prehensile tails. They live to about 14 years in the wild, and in the protection of captivity, up to around 20 years.

The squirrel monkey is the most common of all Amazon monkeys, widespread throughout the lowland rainforest. They roam in large troops of 20 to 100 and make a lot of noise. They feed on small insects, fruit and nectar and particularly like foraging along river edges, so are frequently visible along the Amazon's waterways.

Left: Gray's saki
is a seedeater.

The capuchins have prehensile tails, unlike the squirrel monkeys. They feed on fruit, seeds and insects, roaming through the middle layers of the rainforest in groups of up to 30.

Finally, the only nocturnal primate, the douroucouli, has good night vision through its huge eyes – good enough to run along tree limbs, leaping and performing remarkable acrobatics even on the darkest nights. The night or owl monkey, as it is sometimes called, subsists on a diet of fruit, leaves, insects, flowers and birds' eggs.

Douroucoulis are monogamous and travel in family packs. They are quite small – adults usually weigh about 1kg (2lb) and are about 30–40cm (10–16in) long.

Below: A brown capuchin monkey in the upper-middle forest layer.

Right: A douroucouli – also called the night or owl monkey.

Marmosets

Marmosets and tamarins are the only Amazonian monkeys to have claws instead of nails on their hands and feet, although their great toes have nails. Because of this, they are excellent climbers, using their claws to dig into bark in the same way as squirrels. Although their tails are not prehensile, they are still long and help with balance as the monkeys race through the trees.

The main differences between these two small species is in their teeth. The marmosets have specialised, lower, chisel-like incisors that allow them to gouge holes in the tree bark and then feed upon the resin that flows out.

Left: A marmoset family: the black and white tassel-eared father still carries his almost fully grown twin infants.

Above: Golden white marmoset father with almost fully-grown twins.

Marmosets weigh from 225gm to 453gm (8–16oz); they are around 50cm (22in) long including a 30cm (12in) tail – although the pygmy marmoset is much smaller (see page 56). They live for around 15 years, eating insects mainly, with some fruit and leaves. The over-large, chisel-like incisors are used to extract exudates (gums, resin and sap) from trees.

They communicate by a variety of vocalisations, some of which are too high-pitched for the human ear.

Above: The infants learn from the adults exactly how and what to eat.

Right: All these monkeys frequently groom each other, an activity which helps strengthen the bonds between them.

Overleaf, left: Marmosets maintain contact between each other using very high-pitched, almost bird-like, calls.

Overleaf, right: Six week-old infant marmosets investigate a tree sap hole gouged by their father.

Opposite page: An adult golden-white marmoset investigates a tarantula, which it may well then eat.

Left: Marmosets will visit the forest floor, even though this can bring them into danger from ground predators.

Below: Leaves are lifted slowly as they search for insect prey.

As the smallest living monkey, at 10cm (4–5in) long (excluding the tail), the pygmy marmoset is spectacularly nimble and can leap nearly 5m (16ft). Its coat is brownish, flecked with grey on the head and back, with a banded effect on the tail. The long hair on its face and head look like a mane and this has led to its common Spanish name of leoncito ("little lion").

Pygmy marmosets feed mainly on exudates, gouging oval-shaped, one-inch (2.5cm) holes in tree bark and eating the gums and saps. Protein comes in the form of spiders and insects, although they will occasionally eat fruits.

Pygmy marmosets are monogamous, usually giving birth to non-identical twins. They travel in groups of two to six – the adult pair and its offspring. They sleep in tree holes or vine tangles located near their primary feeding source. They live for around 12–15 years in the wild.

They communicate – as do most marmosets – by using scent markings and calling. They produce high-pitched almost bird-like calls and can emit an ultrasonic scream, that humans cannot hear, to express hostility. Within members of their own family groups and between neighbours of the same species they also use a bizarre range of facial expressions and body postures.

Left: The smallest monkey in the world – the pygmy marmoset.

Left: Pigmy marmosets are used by the Tikuna Indians to catch head lice and nits.

Right: Their local name translates as 'little lion-monkey'.

Below: They gouge holes in trees to feed on sap and resin.

Tamarins

In the same family as marmosets, tamarins are usually slightly bigger – the lion tamarins reaching 25–40cm (13–16in) with a 25–38cm (10–15in) tail. The tamarins are often brightly coloured, with fine, silky coats, tufts, manes and moustaches.

Tamarins are quite common throughout western Amazonia. They are wary and fast-moving, with claws on all except the big toe. Their tails are not prehensile, but act as a balancing aid. They feed mainly on fruits, a diet they augment with nectar, flowers, saps, gums and small animals such as frogs, lizards, insects and spiders.

Right: A white-lipped tamarin peers out from behind a branch.

Each tamarin species (and marmoset species, for that matter) is highly adapted to life in its particular area, which is often one boundaried by rivers. They depend on their territory to provide the specialised diet they require. This explains, in part, why it is rare to see more than one species of marmoset or tamarin in a specific area. The exception to this are the saddle-back and emperor tamarins, who sometimes will share areas.

Left: The midas tamarin is named after its golden-coloured hands.

Above: A midas tamarin feeding on tiny fruits in the lower forest level.

Above: The strange looking ochraceous bare-face tamarin.

Right: A pied barefaced tamarin feeding on gum that surrounds parkia seeds in their pods.

Below: Spix's moustached tamarin using facial expression to communicate. Here, tongue flicking means 'keep away'.

Social Structure

All Amazon monkeys are social animals and maintaining close bonds between individuals of the same family forms an important part of their daily behaviour. Some species stick together in large, cohesive groups of 30 or even more. Woolly monkeys have been counted in groups of up to 40, whereas the spiders tend to number between 10 and 30 individuals. The spider monkeys often split into small satellite units of four or five for foraging; but they remain a permanent part of the larger group and are acutely aware of where the others are at all times.

Above: Two marmosets take it in turn to groom each other.

Right: A pied tamarin sees the sticky gum around the seeds of the parkia. However, tamarins cannot gouge wood for saps and gums like marmosets.

Monkey watching
Jacare Creek

Close to my camp, I filmed one of the most magnificent trees in the forest, the pakira. The crown of this tree usually stands head and shoulders above the rest. When fruiting, it is a fantastic sight with thousands of woody fruit hanging on metre (3ft) long stalks, which, from a distance, look like hundreds of roosting bats. When the pods are mature, they open exposing a dozen or more seeds. To prevent them just falling to the forest floor, before something has a chance to disperse them, the tree produces copious amounts of treacle-like gum that glues them in place. However, there is one species of tamarin able to feed on it.

The pied tamarin will leave the relative safety of the lower forest layer and, like a miniature acrobat, climb down the long, gently swaying stalks to lick up the gum. I could hear the high pitched excited chattering of the group I was filming long before I spotted them, as they came to feast on the same tree at about the same time every afternoon. The tamarin is not interested in the seeds, of course, just the fruit; and it is birds and occasionally the larger monkeys who actually disperse the parkia's seeds.

Above: After two weeks, the gum has almost gone. Soon, the remaining seeds will fall to the floor and be eaten by rodents.

Opposite page: The huge tree's massive limbs dwarf the tiny pied tamarins; a slip could be fatal as the forest floor is some 46m (150ft) below.

Right: Clusters of the pods hang on long stalks from the branches of the highest branches. From a distance, they almost look like roosting bats. When the pods first open, thick, translucent gum is produced to hold the seeds in place.

Both woolly and spider monkeys spend most of their lives in the highest trees, but the males rule woolly society whereas for spider monkeys it's the females that are on top: they generally only tolerate the males at close quarters when they are ready to breed. Both these species give birth to single offspring usually at the start of the rainy season. The pregnancy lasts about seven months and the young will not leave their mother's back permanently until they are well over one year old. Intermittent nursing can carry on until they are over two years of age.

The daily routines of these monkeys are quite similar. Waking just before dawn, they often groom each other for 10 or 15 minutes, then move off to feed. After an especially cold night, spider monkeys hang from lofty exposed branches for five minutes or so, sunbathing in order to absorb the warmth of the new day's sun. In spite of their thick black fur, they still get chilled.

Left: Spider monkeys love nectar; here they suck the sweet liquid from the flowers of a canopy vine called 'tail of the macaw' (above).

Throughout the day, behaviour alternates between feeding, resting, socialising and – especially with infants – playing. One of my best memories is of watching a group of five adult and two juvenile spider monkeys playing a game of tag. For half an hour they chased each other about in the crowns of three adjacent trees. One would slap another and then run away being chased until it was slapped back. I even saw one throw a golf ball-sized fruit at another during this game. Sometimes, in an effort to escape being caught, one would leap for a branch, miss it, and fall 15 or 20m (50–65ft) to another far below. Unruffled, they would then clamber back up and rejoin the fun.

Both woolly and spider monkeys have prehensile, or grasping, muscular tails. This is an effective fifth limb that they use to grip branches, allowing them to reach fruits on lower thin branches that would not support their weight. They also span the gap between two trees using their tail and hands, forming a furry bridge that the infants can cross over.

Spiders and woollys, and indeed many other species, are tremendous leapers but, from my observation, the spider monkey has the greater agility; with its long thin limbs it can out-leap and out-swing all of the others.

Left: A golden white tassel-eared marmoset feeds on the white flesh around the 'eyeball' like guarana seeds.

Monkey watching
Jacare Creek

I was privileged to be, as far as I know, the first person ever to witness and film this event in the Amazon forest. With a camera carefully installed in the back of the birth den, I sat every night for two weeks hidden in a hide on top of a small tower. From just 30cm (12in) away I watched as the female marmoset gave birth to twins. The whole group was there and the male moved next to her as they appeared. She delivered them herself and cleaned them, the male then ate the afterbirth and chewed the umbilical cords off. The babies were smaller than my thumb.

It is an experience that I shall never forget.

Opposite page: It's the father and other members of the group who carry and care for the infants; here they are just twenty-four hours old.

Above: Infant marmosets learn how to gouge holes in trees by watching the adults.

Left: Inside the tree hole the marmoset father moves next to the female to take the one-hour old twins from her.

The small monkeys, marmosets and tamarins, differ from their larger relatives in a number of really fascinating ways. At night when they sleep their body temperature can drop as much as 2°C (4°F). This is one of the reasons why they are extra vulnerable to predators at night because waking up takes longer. I have watched some groups of marmosets appear from their tree hole where they spent the night, more than one hour after sunrise.

What these small primates lack in terms of waking up, they more than make up for in agility and speed of travel. Once on the move they head for the nearest fruiting tree but they also love insects and are constantly on the look-out for them. They will even eat small snakes and poisonous centipedes but their greatest specialisation is their ability to feed on tree sap or resin.

Only marmosets have evolved lower teeth that can chisel into wood. The freshly-gouged hole fills with sap or resin to heal the bark wound and the marmosets feed on it. This nutritious food is very beneficial at times when fruits are scarce. The tamarins do not have these specialised teeth but in some areas have learned how to take advantage of it. In western Brazil, just north of the Amazon

River lives the smallest monkey in the world – the pygmy marmoset. Like its relatives the pygmy marmoset can gouge wood but in this region some tamarins wait until the pygmys have had their fill and then move in to the same tree to lick up the sap that continues to flow from the hole.

Previous page: Golden white tassel-ear marmoset group queuing up to be groomed.

Opposite page, top: Marmosets pass the nights in tree holes or thick tangles of vine.

Opposite page, bottom: Maué's marmoset male with nine-week-old infant.

Above: Golden white marmoset father with almost fully-grown twins.

Left: White marmoset female at a gouge hole site (visible under hand).

Right: An adult female marmoset eating a huge beetle grub.

Below left: Marmosets and tamarins will even eat small snakes.

Below right: Two white marmoset males 'facing off'. They use a lot of body language in order to communicate.

Marmosets and tamarins differ in another remarkable way from their larger relatives where breeding is concerned. Typically, they produce twins and once born the tiny babies are taken from the mother, usually within the first few days of life, and thereafter carried about and cared for by the father and the rest of the family.

Tamarin and marmoset babies grow rapidly and are weaned in just ten to 12 weeks by which time the female is often pregnant again. During their infancy the adults teach the youngsters how to find food, gouge wood for sap and until they are old enough, will share captured insects with them.

As the groups move about they maintain contact with each other using high-pitched almost bird-like calls. They are very territorial, constantly scent-marking their area, and will vigorously defend their patch from neighbouring groups. If interlopers are spotted the group becomes extremely animated. The first impulse is to pull extraordinary faces at each other across a gap between the trees. However, if that doesn't scare them off, they give chase and – just occasionally – aggressive physical contact occurs. In seconds, severe facial wounds can be inflicted on an enemy with those razor-sharp teeth.

Above: A black and white tassel eared marmoset clings securely to a vertical trunk with clawed fingers.

Right: A golden white tassel eared marmoset, alert to every sound and movement.

Above: Unique images: an adult satere male with young twin. This species was only discovered in 1998.

Left: Until marmoset infants are old enough to obtain their own food, they often steal it straight from the mouths of the adults.

Right: Spix's moustached tamarin feeding on sweet nectar from a flowering vine.

Above: A female woolly monkey with her three-month-old baby.

Left: A juvenile woolly monkey peels the bark off a branch to lick the sticky gum.

Right: An adult female black spider monkey opens her mouth threateningly at the intrusion of the camera.

Predators

The larger monkeys have few natural predators, but there is one that causes tremendous alarm whenever it is spotted – the harpy eagle. This massive bird of prey, with a 2m (6.5ft) wing span, is a majestic aerial assassin and is considered to be the world's most powerful eagle. It usually hunts in the highest layer of the forest where it will perch motionless for long periods in the canopy. The harpy eagle also hides in trees at river edges where many creatures are attracted by the water. Once its prey has been selected, speed, power and huge talons are its formidable weapons. Large primates such as the spider or woolly monkey are plucked from lower branches and taken to a high branch where they are simply ripped apart.

Small monkeys, especially marmosets and tamarins, are much more at risk in the middle and lower forest layers from smaller but equally dangerous birds of prey. Unfortunately for them, the smaller raptors are much more numerous than the harpy eagle. Like their larger relatives, the secret of survival for the small primates is seeing the danger before it sees and hits them.

Left: The margay, a nocturnal tree-climbing cat, is very dangerous for small primates such as marmosets and tamarins.

Above: A harpy eagle waits patiently in the canopy; it feeds mainly on monkeys and sloths.

Monkey watching
Rio Abacaxi

I was once exploring an area of forest south of the Amazon bordering the wonderfully named Rio Abacaxi (Pineapple River). Walking quietly with my local guide we spotted a group of around ten white marmosets feeding in a tree just 10m (32ft) in front of us and about 15m (50ft) above the forest floor. We hid behind a large tree trunk and through my binoculars I had great close-up views of them eating the fruits.

Suddenly, one of the marmosets produced a high-pitched scream and in the blink of an eye it appeared that snowballs were falling from the forest above. It was raining marmosets! As they hit the leaf-covered ground they scattered in all directions, so fast it was difficult to take in. A shadow whizzed across the sun-dappled floor and I caught just a glimpse of a dark hawk as it darted through the undergrowth.

High-pitched alarm calls were filling the air but it was impossible to pinpoint their source. Then, slowly, tentatively peeping out from behind many tree trunks, tiny white faces appeared, their eyes scanning the surroundings for the bird. The din did not abate for a good ten minutes.

Above: The jaguarundi hunts mainly on the forest floor in the daytime.

Below: A margay spots a potential victim.

Opposite page: An ocelot, a deadly diurnal cat that can climb.

90

In Amazon monkey society, safety really does lie in numbers. The more individuals there are in a group the more pairs of eyes are available to spot a potential threat. In some species one monkey will actually operate as a lookout while the others safely forage, their concentration focussed on feeding or resting knowing that their lookout is vigilant.

The smaller primates are also at risk from cats, especially the jaguarundi, which spends much of its life up in the trees. As they sometimes sleep in groups, in tree holes, and can be trapped, they are particularly vulnerable to this hunting cat at night. On the rare occasions that the monkeys visit the ground, they have to keep a particular eye out for terrestrial cats and some snakes, particularly boa constrictors. In addition, marmosets and tamarins are also at risk to arboreal snakes, like the emerald tree boa.

The greatest threat to all the Amazon monkeys, however, undoubtedly comes from the destruction and loss of their habitat caused by humans. Man's seemingly insatiable greed for timber and frightening ignorance of how the rainforest ecosystem works means that the long term survival of the forest itself and viable primate populations in Amazonia is seriously in peril.

Above: Man and monkeys: Many forest Indians use the smaller primates both as fashion accessories and as a way of keeping their scalps clean – for the monkeys eat nits, the larvae of headlice.

Conservation

The Amazon rainforest has existed for millions of years, and only in the last 100 years has there been a significant decline. It has been estimated that during the course of the last century we have managed to destroy half of the world's rainforest and, based on current forecasts, we look set to destroy the rest within the next 25 years unless we do something to stop the destruction.

Many of the primates mentioned in this book are threatened by loss of habitat – in itself a death sentence to animals as specialised as these.

On the positive side, eco-tourism has developed over the last 20 years. If the economies of the countries of Amazonia benefit sufficiently from this, then managed tourism may well prove to be one of the few counter-destructive economic forces available in preserving the rainforest. The more people who visit the forests, the more people will become involved in the race to save them.

Above: A golden white tassel-ear marmoset checks out the inside of a tree hole.

Acknowledgements
from Nick Gordon

I have many personal debts of gratitude to individuals and organisations who have helped me over the years. In particular, I would like to thank the following for their valuable assistance and guidance and for taking the time to support me when I most needed it, which was often:

Anthony Rylands, for explaining the latest taxonomy of Amazonian primates.

Rick West, for tarantulas and too many other things to mention.

Nick Peake for providing me with the opportunity to witness and record the environmental cost of human stupidity in many parts of Brazilian Amazonia.

Nigel Blundell for being instrumental in making another book happen.

My assistants: **Gordon Buchanan**, for five exceptional years and the moments that he recorded on film. **Stephen Terry; Ron Kiley; Almir Cavalcante**; and especially **Neil Shaw**, who filled a vital breach in our crew before the contract ink had even died. He helped me bring my jaguar dream to its final conclusion and kept me sane in the process. They all had, to varying degrees, to put up with my obsession to capture the perfect sequence or shot.

Antonieta Sobralino Cavalcante, for being a fantastic field assistant and organiser, for being bitten by countless creatures, for unflinching support over ten years, and therefore for having shared for much of that time the only true danger of our forest, the 'Hook of Holland and her Mad marc'. And, of course, for accepting my proposal of marriage!

Bibliography

Adalardo de Oliviera, Alexander, and Daly, Douglas C.
Florestas do Rio Negro
(Editora Schwarcz Ltda., 2001)

Attenborough, David
Private Life of Plants (BBC Books, 1995)

Attenborough, David
The Life of Birds (BBC Books, 1998)

Bates, Henry Water
The Naturalist on the River Amazons
(John Murray, 1864)

Benad, Gottfried and Hofmockel, Rainer.
History and Perspectives of Muscle Relaxants

Fawcett, Colonel Percy Harrison
Lost Trails, Lost Cities: An Explorer's Narrative (Funk & Wagnalls, 1953)

Gordon, Nick
Tarantulas, Marmosets and Other Stories: an Amazon Diary (Metro, 1997)

Gordon, Nick
Tarantulas, Marmosets: an Amazon Diary (Metro, 1998)

Gordon, Nick
Heart of the Amazon (Metro, 2002)

Goulding, Michael, Smith, Nigel J.H. and Mahar, Dennis J.
Floods of Fortune: Ecology and Economy along the Amazon
(Columbia University Press, 1996)

Henderson, Andes
The Enchanted Canopy
(Fontana/Collins, 1986)

Schomburgk, Richard
Travels in British Guiana (1922)

Schultes, Richard Evans
Where the Gods Reign: Plants and Peoples of the Columbian Amazon
(Synergetic Press, 1988)

Sick, Helmut
Birds in Brazil: A Natural History
(Princeton University Press, 1993)

Silva, Silvestre and Tassara, Helena
Fruit in Brazil (Empresa des Artes, 1996)

Wallace, Alfred R. (Ed.), Richard Spruce
Notes of a Botanist on the Amazon and Andes
(New York; Macmillan and Company, 1908)

Other wildlife titles published by

Evans Mitchell Books

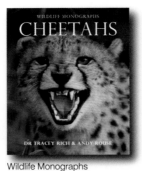

Wildlife Monographs
Cheetahs
ISBN: 978-1-901268-09-6

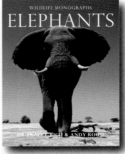

Wildlife Monographs
Elephants
ISBN: 978-1-901268-08-9

Wildlife Monographs
Giant Pandas
ISBN: 978-1-901268-13-3

Wildlife Monographs
Monkeys of the Amazon
ISBN: 978-1-901268-10-2

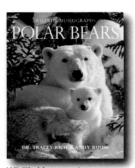

Wildlife Monographs
Polar Bears
ISBN: 978-1-901268-15-7

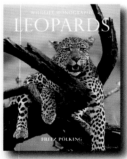

Wildlife Monographs
Loepards
ISBN: 978-1-901268-12-6

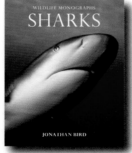

Wildlife Monographs
Sharks
ISBN: 978-1-901268-11-9

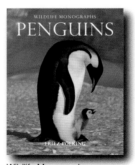

Wildlife Monographs
Penguins
ISBN: 978-1-901268-14-0

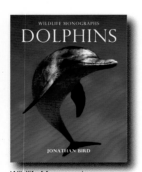

Wildlife Monographs
Dolphins
ISBN: 978-1-901268-17-1

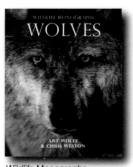

Wildlife Monographs
Wolves
ISBN: 978-1-901268-18-8

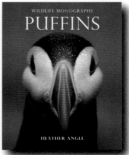

Wildlife Monographs
Puffins
ISBN: 978-1-901268-19-5